Exam Success
IEE Wiring Regulations 2382-20

First published 2008
Reprinted 2008 (three times)

© 2008 The City and Guilds of London Institute
City & Guilds is a trademark of the
City and Guilds of London Institute

Published in association with the Institution of
Engineering and Technology

ISBN: 978-0-86341-886-0

Every effort has been made to ensure that the
information contained in this publication is
true and correct at the time of going to press.
However, examination products and services
are subject to continuous development and
improvement and the right is reserved to
change products and services from time
to time. While the author, publishers and
contributors believe that the information
and guidance given in this work is correct,
all parties must rely upon their own skill and
judgement when making use of it. Neither the
author, the publishers nor any contributor
assume any liability to anyone for any loss or
damage caused by any error or omission in the
work, whether such error or omission is the
result of negligence or any other cause. Where
reference is made to legislation it is not to be
considered as legal advice. Any and all such
liability is disclaimed.

Cover and book design by CDT Design Ltd
Implementation by Adam Hooper, Hoop Design
Typeset in Congress Sans and Gotham
Printed in the UK by Burlington Press

With thanks to Brian Scaddan

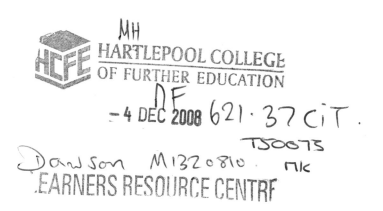

Exam Success
IEE Wiring Regulations 2382-20

City & Guilds Level 3 Certificate in the Requirements for Electrical Installations (16th to 17th Edition Update BS 7671: 2008)

City & Guilds is the UK's leading provider of vocational qualifications, offering over 500 awards across a wide range of industries, and progressing from entry level to the highest levels of professional achievement. With over 8500 centres in 100 countries, City & Guilds is recognised by employers worldwide for providing qualifications that offer proof of the skills they need to get the job done.

Copies may be obtained from:
Teaching & Learning Materials
City & Guilds
1 Giltspur Street
London EC1A 9DD
For publications enquiries:
T +44 (0)20 7294 4113
F +44 (0)20 7294 3414
Email learningmaterials@cityandguilds.com

The Institution of Engineering and Technology is the new institution formed by the joining together of the IEE (The Institution of Electrical Engineers) and the IIE (The Institution of Incorporated Engineers). The new institution is the inheritor of the IEE brand and all its products and services including the IEE Wiring Regulations (BS 7671) and supporting material.

Copies may be obtained from:
The Institution of Engineering and Technology
P.O. Box 96
Stevenage
SG1 2SD, UK
T +44 (0)1438 767 328
Email sales@theiet.org
www.theiet.org

Contents

Introduction   6

The exam

The exam   10

Sitting a City & Guilds online examination   11

Frequently asked questions   17

Exam content   19

Tips from the examiner   24

Exam practice 1

Sample test 1   28

Questions and answers   35

Answer key   44

Exam practice 2

Sample test 2   46

Questions and answers   53

Answer key   62

More information

Further reading   64

Online resources   65

Further courses   66

Introduction

How to use this book

This book has been written as a study aid for the City & Guilds Level 3 Certificate in the Requirements for Electrical Installations (16th to 17th Edition Update BS 7671: 2008) (2382-20). This update certificate is designed for those candidates who have already achieved their Level 3 Certificate in the Requirements for Electrical Installations BS 7671: 2001 (2381) and who need to prove that they are conversant with the updates and changes introduced in the Seventeenth Edition of the *IEE Wiring Regulations*.

This book sets out methods of studying, offers advice on exam preparation and provides details of the scope and structure of the examination, alongside two sample exam papers with answers and references to the relevant section of the *IEE Wiring Regulations*. Used as a study guide for exam preparation and practice, it will help you to reinforce and test your existing knowledge, and will give you guidelines and advice about sitting the exam. You should try to answer the sample test questions under exam conditions (or as close as you can get) and then review all of your answers. This will help you to become familiar with the types of question that might be asked in the exam and will also give you an idea of how to pace yourself in order to complete all questions comfortably within the time limit. This book cannot guarantee a positive exam result, but it can play an important role in your overall revision programme, enabling you to focus your preparation and approach the exam with confidence.

City & Guilds Level 3 Certificate in the Requirements for Electrical Installations (16th to 17th Edition Update BS 7671: 2008)

If you are a practising electrician with relevant experience and you are already in possession of your 2381 Certificate based on the Sixteenth Edition of the *IEE Wiring Regulations*, you will also need to achieve this Level 3 update Certificate to show you have a sound working knowledge of the format, content and application of the current edition of the wiring regulations, the *IEE Wiring Regulations Seventeenth Edition* (BS 7671: 2008).

The syllabus covers the new material introduced in the Seventeenth Edition. Topics include protection for safety, selection and erection of equipment, and inspection and testing to meet the standards of the *IEE Wiring Regulations*. If successful, you could go on to take other City & Guilds electrical qualifications, such as Inspection, Testing and Certification of Electrical Installations (2391-10), and In-service Inspection and Testing of Electrical Equipment (2377).

IEE Wiring Regulations Seventeenth Edition

You will need a copy of the *IEE Wiring Regulations Seventeenth Edition* to be able to answer the sample questions and in order to revise for the examination. The *IEE Wiring Regulations Seventeenth Edition*, also called *BS 7671: 2008 Requirements for Electrical Installations*, is the national standard for the electrical industry in respect of safe use and operation of electrical equipment and systems in the United Kingdom, in accordance with the recommendations of the Institution of Engineering and Technology. It contains rules for the design and erection of electrical installations, so as to provide for safety and proper functioning for the intended use.

Finding a centre

In order to take the exam, you must register at an approved City & Guilds centre. You can find your nearest centre by looking up the qualification number 2382-20 on www.cityandguilds.com. The IET is an accredited centre and runs online exams in different parts of the country. For more details, see www.theiet.org.

At each centre, the Local Examinations Secretary will enter you for the award, collect your fees, arrange for your assessment to take place and correspond with City & Guilds on your behalf. The Local Examinations Secretary also receives all of your certificates and correspondence from City & Guilds. Most centres will require you to attend a course of learning before entering you for the examination. These are usually one- or two-day courses, but can also run across a term, once a week.

Awarding and reporting

When you complete the City & Guilds 2382-20 Certificate online examination, you will be given your provisional results, as well as a breakdown of your performance in the various areas of the examination. This is a useful diagnostic tool if you fail the exam, as it enables you to identify your individual strengths and weaknesses across the different topics.

A Certificate is issued automatically when you have been successful in the assessment, but it will not indicate a grade or percentage pass. Your centre will receive your Notification of Candidate's Results and Certificate. Any correspondence is conducted through the centre. The centre will also receive consolidated results lists detailing the performance of all candidates entered. If you have particular requirements that will affect your ability to attend and take the examination, then your centre should refer to City & Guilds policy document 'Access to Assessment: Candidates with Particular Requirements'.

Notes

The exam

The exam	10
Sitting a City & Guilds online examination	11
Frequently asked questions	17
Exam content	19
Tips from the examiner	24

The exam

The exam

The examination has a multiple-choice format, with 30 questions, which you will have one hour to answer. The test is offered on GOLA, a simple online service that does not require strong IT skills. GOLA uses a bank of questions set and approved by City & Guilds. Each candidate receives randomised questions, so no two candidates will sit exactly the same test.

The exam follows a set structure, based on the seven parts of the *IEE Wiring Regulations* plus the appendices. The table below outlines the sections of the exam and the number of questions in each section. It also shows the weighting – so you can see how important each section is in determining your final score.

Section	Topic		% weighting	No of questions
1	Part 1	Scope, object and fundamental principles	7	2
2	Part 2	Definitions	7	2
3	Part 3	Assessment of general characteristics	3	1
4	Part 4	Protection for safety	26	8
5	Part 5	Selection and erection of equipment	20	6
6	Part 6	Inspection and testing	7	2
7	Part 7	Special installations or locations	23	7
8	Use of appendices		7	2
	Total		100	30

Sitting a City & Guilds online examination

The test will be taken under usual exam conditions. You will not be able to refer to any materials or publications other than the ones that are approved for this test (the Seventeenth Edition of the *IEE Wiring Regulations*). You will not be allowed to take your mobile phone into the exam room and you cannot leave the exam room unless you are accompanied by one of the test invigilators. If you leave the exam room unaccompanied before the end of the test period, you may not be allowed to come back into the exam.

When you take a City & Guilds test online, you can go through a tutorial to familiarise yourself with the online procedures. When you are logged on to take the exam, the first screen will give you the chance to go into a tutorial. The tutorial shows how the exam will be presented and how to get help, how to move between different screens, and how to mark questions that you want to return to later.

City& Guilds

Please work through the tutorial before you start your examination.

This will show you how to answer questions and use the menu options to help you complete the examination.

Please note that examination conditions now apply.

The time allowed for the tutorial is 10 minutes.

Click on Continue to start the tutorial or Skip to go straight to the examination.

| Skip | Continue |

Notes

The sample questions in the tutorial are unrelated to the exam you are taking. The tutorial will take 10 minutes, and is not included in the test time. The test will only start once you have completed or skipped the tutorial. A screen will appear that gives the exam information (the time, number of questions and name of the exam).

City& Guilds

Examination: 2382-20 Requirements for Electrical Installations Update

Number of questions: 30

Time allowed: 60 minutes

Note: Examination conditions now apply.

The next screen that will appear is the Help screen, which will give you instructions on how to navigate through this examination. Please click OK to view the Help screen.

The time allowed for the examination will start after you have left the Help screen.

A warning message will appear 5 minutes before the end of the examination.

OK

After clicking 'OK', the Help screen will appear. Clicking the 'Help' button on the tool bar at any time during the exam will recall this screen.

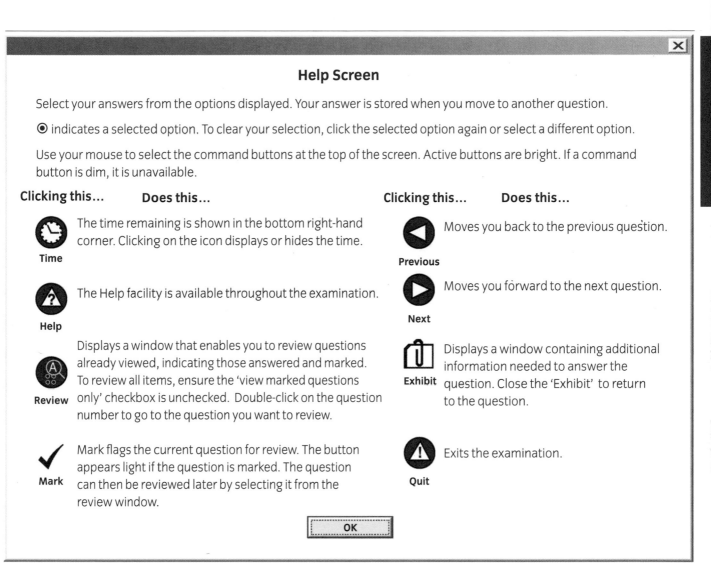

Help Screen

Select your answers from the options displayed. Your answer is stored when you move to another question.

⊙ indicates a selected option. To clear your selection, click the selected option again or select a different option.

Use your mouse to select the command buttons at the top of the screen. Active buttons are bright. If a command button is dim, it is unavailable.

Clicking this…	Does this…	Clicking this…	Does this…
Time	The time remaining is shown in the bottom right-hand corner. Clicking on the icon displays or hides the time.	**Previous**	Moves you back to the previous question.
Help	The Help facility is available throughout the examination.	**Next**	Moves you forward to the next question.
Review	Displays a window that enables you to review questions already viewed, indicating those answered and marked. To review all items, ensure the 'view marked questions only' checkbox is unchecked. Double-click on the question number to go to the question you want to review.	**Exhibit**	Displays a window containing additional information needed to answer the question. Close the 'Exhibit' to return to the question.
Mark	Mark flags the current question for review. The button appears light if the question is marked. The question can then be reviewed later by selecting it from the review window.	**Quit**	Exits the examination.

OK

After clicking 'OK' while in the Help screen, the exam timer will start and you will see the first question. The question number is always shown in the lower left-hand corner of the screen. If you answer a question but wish to return to it later, then you can click the 'Flag' button. When you get to the end of the test, you can choose to review these flagged questions.

Notes

Protection against electric shock under single-fault conditions is defined as

○ a overload protection

○ b fault protection

○ c basic protection

○ d undervoltage protection.

Question Number 1

If you select 'Quit' on the tool bar at any point, you will be given the choice of ending the test. **If you select 'Yes', you will not be able to go back to your test.**

If you click 'Time' on the tool bar at any point, the time that you have left will appear in the bottom right-hand corner. When the exam timer counts down to five minutes remaining, a warning will flash on to the screen.

Some of the questions in the test may be accompanied by pictures. The question will tell you whether you will need to click on the 'Exhibit' button to view an image.

When you reach the final question and click 'Next', you will reach a screen that allows you to 'Review your answers' or 'Continue' to end the test. You can review all of your answers or only the ones you have flagged. To review all your answers, make sure that the 'view marked questions only' checkbox is unchecked (click to uncheck). After you have completed your review, you can click 'Continue' to end the test.

Notes

City& Guilds

You have answered 30   questions out of a total of 30

To check your answers and return to the examination, click on the Review button. If your time has expired, you cannot return to the examination.

If you wish to submit your answers and end the examination, please click the Continue button.

Clicking Continue will end the examination.

| Review | Continue |

Once you choose to end the exam by clicking 'Continue', the 'Test completed' screen will appear. Click on 'OK' to end the exam.

At the end of the exam, you will be given an 'Examination Score Report'. This gives a provisional grade (pass or fail) and breakdown of score by section. This shows your performance in a bar chart and in percentage terms, which allows you to assess your own strengths and weaknesses. If you did not pass, it gives valuable feedback on which areas of the course you should revise before re-sitting the exam.

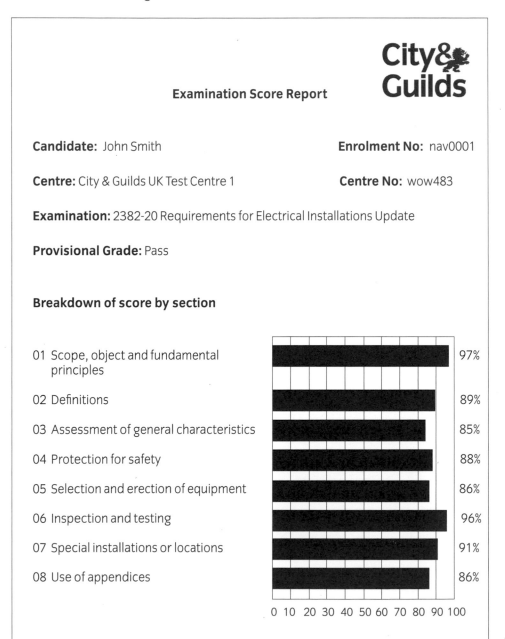

City& Guilds

Examination Score Report

Candidate: John Smith   **Enrolment No:** nav0001

Centre: City & Guilds UK Test Centre 1   **Centre No:** wow483

Examination: 2382-20 Requirements for Electrical Installations Update

Provisional Grade: Pass

Breakdown of score by section

01 Scope, object and fundamental principles	97%
02 Definitions	89%
03 Assessment of general characteristics	85%
04 Protection for safety	88%
05 Selection and erection of equipment	86%
06 Inspection and testing	96%
07 Special installations or locations	91%
08 Use of appendices	86%

0 10 20 30 40 50 60 70 80 90 100

This chart provides feedback to show candidate performance for each section of the test.
It should be used along with the Test Specification, which can be found in the Scheme Handbook.

Frequently asked questions

When can I sit the paper?

You can sit the exam at any time, as there are no set exam dates. You may need to check with your centre when it is able to hold exam sessions.

Can I use reference books in the test?

Yes, you can use a copy of the IEE Wiring Regulations Seventeenth Edition.

How many different parts of the test are there?

There are 30 questions, which cover eight different sections of the syllabus.

Do I have a time limit for taking the test?

You have one hour to complete the test.

Do I need to be good at IT to do the test online?

No, the system is really easy to use, and you can practise before doing the test. There is also a practice GOLA test available to try on the IET website.

What happens if the computer crashes in the middle of my test?

This is unlikely, because of the way the system has been designed. If there is some kind of power or system failure, then your answers will be saved and you can continue on another machine if necessary.

Can people hack into the system and cheat?

There are lots of levels of security built into the system to ensure its safety. Also, each person gets a different set of questions, which makes it very difficult to cheat.

Can I change my answer?

Yes, you can change your answers quickly, easily and clearly at any time in the test up to the point where you end the exam. With any answers you feel less confident about, you can click the 'Flag' button, which means you can review these questions before you end the test.

How do I know how long I've got left to complete the test?

You can check the time remaining at any point during the exam by clicking on the 'Clock' icon in the tool bar. The time remaining will come up on the bottom right corner of the screen.

Is there only one correct a, b, c or d answer to multiple-choice questions?

Yes.

What happens if I don't answer all of the questions?

You should attempt to answer all of the questions. If you find a question difficult, mark it using the 'Flag' button and return to it later.

What grades of pass are there?

A Pass or a Fail.

When can I resit the test if I fail?

You can resit the exam at any time, and as soon as you and your tutor decide it is right for you, subject to the availability of the online examination.

Exam content

To help you to fully understand the exam content, this chapter is divided into the eight sections of the exam, which are in turn mapped to the seven parts and the appendices of BS 7671 (the *IEE Wiring Regulations Seventeenth Edition*). The requirements of the examination are listed and references made to the relevant parts of BS 7671. One of the most useful parts of BS 7671 is its very comprehensive index. If you don't know where to look, use the index.

Section 1

Part 1 – Scope, object and fundamental principles

There are two questions on Part 1, which represent 7 per cent of the mark. Part 1 of BS 7671 contains Chapters 11, 12 and 13 and is the essence of the standard. It includes the scope, an important section as any installation for which requirements are being sought must be confirmed as being within the scope of BS 7671. It also includes the fundamental principles (Chapter 13). The rest of BS 7671, that is Parts 2 to 7, supports Part 1; Part 2 provides definitions of the terminology used throughout the standard and Parts 3 to 7 provide technical requirements intended to ensure that the electrical installation conforms with the fundamental principles of Chapter 13 of Part 1.

It is to be noted that departures from Parts 3 to 7 are allowed (Regulations 120.3 and 120.4), providing there is still compliance with Part 1. This allows the use of new materials and inventions not anticipated when the Regulations were published. However, departures from Parts 3 to 7, including the use of new materials and inventions, must not result in a degree of safety any less than that required by Part 1. Departures from Parts 3 to 7 must be recorded on the installation certificate.

To show that you are conversant with BS 7671, you are required to be able to:

1.1 Identify examples of installations in the scope of BS 7671 and particular requirements for specific installations and locations (for Scope, see Chapter 11 and Regulation 110.1)

1.2 Identify those installations that are excluded from BS 7671 (see Chapter 11 and Regulation 110.2)

1.3 Identify requirements for protection against voltage disturbances (see Regulation 131.6)

1.4 Identify the measures required for protection from electromagnetic influences (see Regulation 131.6).

Section 2

Part 2 – Definitions

There are two questions on Part 2, which represent 7 per cent of the mark. There are approximately 12 pages dedicated to the definitions used by the Regulations. Definitions that are particularly important, as their meanings in BS 7671 are often not the same as those in common English usage, include:

> **Cold tail.** The interface between the fixed installation and a heating unit.
>
> **Line conductor.** A conductor of an a.c. system for the transmission of electrical energy other than a neutral conductor, a protective conductor or a PEN conductor.
>
> **Protective earthing.** Earthing a point or points in a system or in an installation or in equipment for the purposes of safety.

To show that you are conversant with BS 7671, you are required to be able to:
2.1 Use Part 2 of BS 7671.

Section 3

Part 3 – Assessment of general characteristics

There is one question on Part 3, which represents 3 per cent of the mark. Part 3 requires a general assessment of the characteristics of the proposed installation. It also requires that the nature or characteristics of the supply be determined, as this plays an important part in the design of the installation. Consideration needs to be given to safety services and standby supplies. Part 3 also includes requirements for the installation circuit arrangements.

To show that you are conversant with BS 7671, you are required to be able to:
3.1   State the source (eg standby, external) and characteristics necessary for a supply (see Section 331)
3.2   State the need to divide an installation into suitable circuit arrangements (see Chapter 31, Section 314)
3.3   State the need for an assessment of each circuit regarding continuity of service (see Chapter 36).

Section 4

Part 4 – Protection for safety

There are eight questions on Part 4, which represent 26 per cent of the mark. Part 4 contains some of the key chapters of BS 7671: Chapter 41 Protection against electric shock, Chapter 42 Protection against thermal effects and Chapter 43 Protection against overcurrents.

To show that you are conversant with BS 7671, you are required to be able to: **Notes**

4.1   Identify the differences between basic and fault protection
          (see Part 2 Definitions)
4.2   State means of protection against electric shock by
          a   basic protection (see Sections 416 and 417)
          b   fault protection (see Sections 412, 413, 414 and 418)
          c   both basic and fault protection (excluding IT) (see Regulations
               412.2 and 414.2)
          d   additional protection (see Section 415)
4.3   Describe how the requirements for shock protection are affected by
          a   the value of the external loop impedance (Z_e)
          b   compliance with $Z_s = Z_e + (R_1 + R_2)$
          c   compliance with Tables 41.1, 41.2, 41.3, 41.4, 41.5 and 41.6
               (see Part 2, Symbols used in the Regulations)
4.4   Describe means of protection against fire, burns and harmful thermal
          effects and identify precautions where particular risks of danger of fire
          exist (see Chapter 42, Section 422)
4.5   State the requirements for protection against
          a   voltage disturbances
               (i)   overvoltage (see Chapter 44, Section 443)
               (ii)   undervoltage (see Chapter 44, Section 445)
          b   electromagnetic disturbances.

Section 5

Part 5 – Selection and erection of equipment

There are six questions on Part 5, which represent 20 per cent of the mark.
Part 5 includes requirements for busbar trunking and powertrack wiring
systems, monitoring devices, earthing, RCDs and generating sets.

To show that you are conversant with BS 7671, you are required to be able to:
5.1   Identify the requirements for selection and erection of busbar trunking
          and powertrack wiring systems (see Regulation 521.4)   .
5.2   Identify requirements for cables concealed in a floor or ceiling
          (see Regulations 522.6.4 and 522.6.5)
5.3   Identify devices used for protection against:
          a   the risk of fire (see Section 532)
          b   overcurrent (see Section 533)
5.4   Describe the function of, and the devices used for, monitoring:
          a   insulation (see Regulations 538.1, 538.2 and 538.3)
          b   residual current (see Regulation 538.4)
5.5   Identify under what circumstances metal water pipes may be used as
          earth electrodes (see Regulation 542.2)

5.6 State the requirements for the protection of socket-outlets by a residual current device (see Regulations 411.3.3 (i) and 415.1)

5.7 Identify the requirements for luminaires and lighting installations (see Section 559)

5.8 State the requirements for low voltage generating sets (see Section 551)

5.9 State that cables are subject to electromechanical and electromagnetic stress in addition to thermal damage under fault conditions (see Regulations 521.5.1 and 521.5.2)

5.10 State the requirements for supplies for safety services and their associated circuits and cables (see Chapter 56).

Section 6

Part 6 – Inspection and testing

There are two questions on Part 6, which represent 7 per cent of the mark.

To show that you are conversant with BS 7671, you are required to be able to:

6.1 State the requirements for protection by SELV, PELV or by electrical separation (see Regulation 612.4)

6.2 State minimum values of insulation resistance (see Regulation 612.3)

6.3 State the requirements for verification of phase sequence (see Regulation 612.12)

6.4 State the requirements for verification of voltage drop (see Regulation 612.14).

Section 7

Part 7 – Special installations or locations

There are usually seven questions on Part 7, which represent 23 per cent of the mark. Part 7 includes additional requirements or altered requirements to those in Parts 3 to 5. The requirements of the rest of BS 7671 apply to the special locations, unless they are amended or altered by the particular requirements in Part 7.

To show that you are conversant with BS 7671, you are required to be able to:

7.1 State the requirements for safety measures in a location containing a bath or shower (see Section 701)

7.2 State the special precautions that must be applied regarding swimming pools and other basins (see Section 702)

7.3 State the requirements for safety measures in a room or cabin containing sauna heaters (see Section 703)

7.4 Identify the requirements relevant to construction and demolition site installations (see Section 704)

7.5 Identify the requirements relevant to installations within agricultural and horticultural premises (see Section 705)

7.6 Identify the requirements for electrical installations in caravans, motor caravans and caravan parks (see Sections 708 and 721)

7.7 Identify the requirements for conducting locations with restricted movement (see Section 706)

7.8 Identify the requirements for marinas and similar locations (see Section 709)

7.9 Identify the requirements for exhibitions, shows and stands (see Section 711)

7.10 Identify the requirements for locations containing solar photovoltaic power supply systems (see Section 712)

7.11 Identify the requirements for mobile or transportable units (see Section 717)

7.12 Identify the requirements for temporary electrical installations for structures, amusement devices and booths at fairgrounds, amusement parks and circuses (see Section 740)

7.13 Identify the requirements for locations containing floor and ceiling heating systems (see Section 753).

Section 8

There are two questions on the Appendices of BS 7671, which represent 7 per cent of the mark.

To show that you are conversant with BS 7671, you are required to be able to:

8.1 Apply relevant information/data within Appendices.

Notes

Notes

Tips from the examiner

The following tips are intended to aid confident test performance. Some are more general and would apply to most exams. Others are more specific, either because of the format of this test (multiple choice) or the nature of the subject.

✔ If you rarely use a computer, try to get some practice beforehand. You need to be able to use a mouse to move a cursor arrow around a computer screen, as you will use the cursor to click on the correct answer in the exam.

✔ Take a copy of the *IEE Wiring Regulations Seventeenth Edition* (*BS 7671: 2008 Requirements for Electrical Installations*) into the exam. Take the time to familiarise yourself with the structure and content of this publication.

✔ Make the most of the course you will attend before taking the exam. Try to attend all sessions and be prepared to devote time outside the class to revise for the exam.

✔ On the day of the exam, allow plenty of time for travel to the centre and arrive at the place of the exam at least 10 minutes before it's due to start so that you have time to relax and get into the right frame of mind.

✔ Listen carefully to the instructions given by the invigilator.

✔ Read the question and every answer before making your selection. Do not rush – there should be plenty of time to answer all the questions.

✔ Look at the exhibits where instructed. Remember, an exhibit supplies you with information that is required to answer the question.

✔ Attempt to answer all the questions. If a question is not answered, it is marked as wrong. Making an educated guess improves your chances of choosing the correct answer. Remember, if you don't select an answer, you will definitely get no marks for that question.

✔ The order of the exam questions follows the order of the *IEE Wiring Regulations Seventeenth Edition* publication. Therefore, look for the answers to early questions at the front of the book and progress through it as you work through the exam questions.

Notes

✔ Don't worry about answering the questions in the order in which they appear in the exam. Choose the 'Flag' option on the tool bar to annotate the questions you want to come back to. If you spend too much time on questions early on, you may not have time to answer the later questions, even though you know the answers.

✔ Although not absolutely necessary, some candidates find it useful to bring a basic, non-programmable calculator with square-root ($\sqrt{}$) and square (x^2) functions to the exam.

✔ If you are having trouble finding the regulation for a particular question, look for the subject in the index of the *IEE Wiring Regulations Seventeenth Edition*. Using the index should minimise the time it takes you to find the relevant topic.

✔ It is not recommended that you memorise any of the material presented here in the hope it will come up in the exam. The exam questions featured in this book will help you to gauge the kinds of questions that might be asked. It is highly unlikely you will be asked any identical questions in the exam, but you may see variations on certain themes.

Notes

Exam practice 1

Sample test 1	28
Questions and answers	35
Answer key	44

Exam practice 1

Sample test 1

The sample test below has 30 questions, the same number as the online exam, and its structure follows that of the online exam. The test appears first without answers, so you can use it as a mock exam. It is then repeated with answers and references to the relevant section of the *IEE Wiring Regulations Seventeenth Edition*. Finally, there is an answer key for easy reference.

Answer the questions by filling in the circle next to your chosen option.

Section 1

1 BS 7671 applies to

○ a   lift installations
○ b   highway equipment
○ c   equipment on board ships
○ d   electrical equipment of machines.

2 Where protection for persons and livestock against injury and against damage to property is required, which of the following need not be taken into account?

○ a   Overvoltage due to switching
○ b   Undervoltage and subsequent voltage recovery
○ c   Direct lightning strikes
○ d   Faults between live conductors supplied at different voltages

Section 2

3 Protection against electric shock under single-fault conditions is defined as

○ a   overload protection
○ b   fault protection
○ c   basic protection
○ d   undervoltage protection.

4 The symbol for rated current of a protective device is

- ○ a I_b
- ○ b I_n
- ○ c I_t
- ○ d I_a.

Section 3

5 Which of the following supply characteristics would need to be ascertained for a new domestic installation?

- ○ a Number of points of utilization
- ○ b The supply transformer type
- ○ c The supply cable size
- ○ d The nature of the current and frequency

Section 4

6 Protection by automatic disconnection of supply is

- ○ a permitted only if the installation is under effective supervision
- ○ b a method of reducing magnetic effects
- ○ c a combination of basic and fault protection
- ○ d a combination of thermal and overvoltage protection.

7 When calculating earth fault loop impedance, U_o represents the

- ○ a design current
- ○ b nominal line voltage to Earth
- ○ c prospective short-circuit current
- ○ d open circuit voltage of the supply transformer.

8 A 6A BS EN 61009 RCBO with a maximum value of earth fault loop impedance of 1.92 Ω is type

- ○ a A
- ○ b B
- ○ c C
- ○ d D.

9 **The maximum disconnection time for a circuit supplied by a reduced low voltage system using a 110 V midpoint earthed transformer is**

○ a   0.2 second
○ b   0.4 second
○ c   1 second
○ d   5 seconds.

10 **Where basic protection and/or fault protection is provided, certain external influences may require additional protection provided by**

○ a   obstacles
○ b   placing out of reach
○ c   the use of 30 mA RCDs
○ d   the use of time delayed 100 mA RCDs.

11 **The horizontal top surface of a barrier or enclosure which is readily accessible shall provide a degree of protection of at least**

○ a   IP55 or IP66
○ b   IPX4 or IPXX7
○ c   IPXXB or IP2X
○ d   IPXXD or IP4X.

12 **Additional protection against shock is provided by**

○ a   BS 3036 fuses
○ b   BS EN 60898 circuit breakers
○ c   time delayed 100 mA RCDs
○ d   30 mA RCDs.

13 **An undervoltage device has operated and restoring the supply may cause danger. The reclosure of this device should be**

○ a   automatic when under the supervision of a competent person
○ b   manually operated
○ c   possible only with the use of a key or tool
○ d   automatic with time delay.

Section 5

14 Where underground power and telecommunication cables are in close proximity, they should be separated by a minimum distance of

- ○ a   1000 mm
- ○ b   500 mm
- ○ c   300 mm
- ○ d   100 mm.

15 An RCD that is installed for protection against the risk of fire where combustible materials are stored shall be

- ○ a   installed at the farthest point of the circuit
- ○ b   installed at the origin of the circuit
- ○ c   arranged to switch line conductors
- ○ d   rated at 500 mA.

16 Where a 30 mA RCD is installed upstream of a residual current monitor (RCM), the rating of the RCM shall not exceed

- ○ a   3 mA
- ○ b   10 mA
- ○ c   30 mA
- ○ d   100 mA.

17 Which of the following may be used as a protective conductor?

- ○ a   An oil service pipe
- ○ b   A gas service pipe
- ○ c   Flexible or pliable conduit
- ○ d   An extraneous-conductive-part

18 Where a step-up transformer is used, all live conductors of the supply shall be disconnected from the supply by a

- ○ a   single-pole switch
- ○ b   two-way switch
- ○ c   linked switch
- ○ d   fused switch.

Notes

19 Which of the following is the designated symbol for a high pressure sodium lamp with an internal starting device?

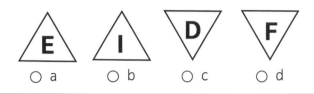

○ a   ○ b   ○ c   ○ d

Section 6

20 Prospective fault current is recorded using the highest value determined from the prospective

○ a   short circuits and earth leakage currents
○ b   short circuit and earth fault currents
○ c   earth leakage and breaking capacity currents
○ d   breaking capacity and earth fault currents.

21 For SELV and PELV circuits the separation of live parts from those of other circuits must be confirmed by

○ a   inspection
○ b   measurement
○ c   calculation
○ d   enquiry.

Section 7

22 Local supplementary bonding in a bathroom may be connected

○ a   near the origin of the installation at the consumer unit
○ b   close to the entry of extraneous-conductive-parts to the room
○ c   to non-metallic water service pipes
○ d   to ceramic bath or shower basins.

23 In zones 0 and 1 of a fountain, the protective measure of SELV may be used, provided that the source of SELV is

○ a   inside these zones
○ b   outside these zones
○ c   via an autotransformer
○ d   centre tapped to earth.

24 Which of the following supply systems is <u>not</u> to be used on construction sites, with the exception of a fixed building on the site?

- a   TN-C-S
- b   TT
- c   TN-S
- d   Reduced low voltage

25 Enclosures used to protect electrical equipment vulnerable to external influences in a horticultural installation shall have a minimum degree of protection of

- a   IP22
- b   IP33
- c   IP44
- d   IP55.

26 The maximum length of a flexible cord or cable for a caravan connection is

- a   1.8 m
- b   6 m
- c   10 m
- d   25 m.

27 Any cable intended to supply temporary exhibition structures shall have, at its origin, an RCD having a maximum rated residual operating current of

- a   30 mA
- b   100 mA
- c   300 mA
- d   500 mA.

28 Which of the following is regarded as a mobile or transportable unit?

- a   Mobile generating sets
- b   Mobile medical services unit
- c   Motor caravan
- d   Mobile machinery

Notes

Notes

Section 8

29 The installation reference method for single-core non-sheathed cables in trunking run horizontally on a wooden wall is

- ○ a   A
- ○ b   B
- ○ c   C
- ○ d   E.

30 The rating factor for a 70 °C thermoplastic cable installed in air in an ambient temperature of 40 °C is

- ○ a   1.0
- ○ b   0.91
- ○ c   0.87
- ○ d   0.85.

Questions and answers

The questions and answers in sample test 1 are repeated below with answers and references to the relevant section of the *IEE Wiring Regulations Seventeenth Edition*.

Section 1

1 BS 7671 applies to

○ a   lift installations
◉ b   highway equipment
○ c   equipment on board ships
○ d   electrical equipment of machines.

Answer b
See Part 1: Scope, Regulation 110.1.

2 Where protection for persons and livestock against injury and against damage to property is required, which of the following need <u>not</u> be taken into account?

○ a   Overvoltage due to switching
○ b   Undervoltage and subsequent voltage recovery
◉ c   Direct lightning strikes
○ d   Faults between live conductors supplied at different voltages

Answer c
See Part 1: Scope, Regulations 131.6.1 to 131.6.4.

Section 2

3 Protection against electric shock under single-fault conditions is defined as

○ a   overload protection
◉ b   fault protection
○ c   basic protection
○ d   undervoltage protection.

Answer b
See Part 2: Definitions.

4   The symbol for rated current of a protective device is

- ○ a I_b
- ◉ b I_n
- ○ c I_t
- ○ d I_a.

Answer b
See Part 2: Definitions, Symbols.

Section 3

5   Which of the following supply characteristics would need to be ascertained for a new domestic installation?

- ○ a   Number of points of utilization
- ○ b   The supply transformer type
- ○ c   The supply cable size
- ◉ d   The nature of the current and frequency

Answer d
See Part 3: Assessment of general characteristics, Regulation 313.1.

Section 4

6   Protection by automatic disconnection of supply is

- ○ a   permitted only if the installation is under effective supervision
- ○ b   a method of reducing magnetic effects
- ◉ c   a combination of basic and fault protection
- ○ d   a combination of thermal and overvoltage protection.

Answer c
See Part 4: Protection for safety, Regulation 411.1.

7   When calculating earth fault loop impedance, U_o represents the

- ○ a   design current
- ◉ b   nominal line voltage to Earth
- ○ c   prospective short-circuit current
- ○ d   open circuit voltage of the supply transformer.

Answer b
See Part 4: Protection for safety, Regulation 411.4.5.

8 A 6A BS EN 61009 RCBO with a maximum value of earth fault loop impedance of 1.92 Ω is type

- a   A
- b   B
- c   C
- ◉ d   D.

Answer d
See Part 4: Protection for safety, Regulation 411.4.7, Table 41.3.

9 The maximum disconnection time for a circuit supplied by a reduced low voltage system using a 110 V midpoint earthed transformer is

- a   0.2 second
- b   0.4 second
- c   1 second
- ◉ d   5 seconds.

Answer d
See Part 4: Protection for safety, Regulation 411.8.3.

10 Where basic protection and/or fault protection is provided, certain external influences may require additional protection provided by

- a   obstacles
- b   placing out of reach
- ◉ c   the use of 30 mA RCDs
- d   the use of time delayed 100 mA RCDs.

Answer c
See Part 4: Protection for safety, Regulation 415.1.1.

11 The horizontal top surface of a barrier or enclosure which is readily accessible shall provide a degree of protection of at least

- a   IP55 or IP66
- b   IPX4 or IPXX7
- c   IPXXB or IP2X
- ◉ d   IPXXD or IP4X.

Answer d
See Part 4: Protection for safety, Regulation 416.2.2.

Notes

12 Additional protection against shock is provided by

- ○ a   BS 3036 fuses
- ○ b   BS EN 60898 circuit breakers
- ○ c   time delayed 100 mA RCDs
- ◉ d   30 mA RCDs.

Answer d
See Part 4: Protection for safety, Regulation 415.1.1.

13 An undervoltage device has operated and restoring the supply may cause danger. The reclosure of this device should be

- ○ a   automatic when under the supervision of a competent person
- ◉ b   manually operated
- ○ c   possible only with the use of a key or tool
- ○ d   automatic with time delay.

Answer b
See Part 4: Protection for safety, Regulation 445.1.5.

Section 5

14 Where underground power and telecommunication cables are in close proximity, they should be separated by a minimum distance of

- ○ a   1000 mm
- ○ b   500 mm
- ○ c   300 mm
- ◉ d   100 mm.

Answer d
See Part 5: Selection and erection of equipment, Regulation 528.2.

15 **An RCD that is installed for protection against the risk of fire where combustible materials are stored shall be**

- ○ a installed at the farthest point of the circuit
- ◉ b installed at the origin of the circuit
- ○ c arranged to switch line conductors
- ○ d rated at 500 mA.

Answer b

See Part 5: Selection and erection of equipment, Regulation 532.1.

16 **Where a 30 mA RCD is installed upstream of a residual current monitor (RCM), the rating of the RCM shall not exceed**

- ○ a 3 mA
- ◉ b 10 mA
- ○ c 30 mA
- ○ d 100 mA.

Answer b

See Part 5: Selection and erection of equipment, Regulation 538.4.1.

17 **Which of the following may be used as a protective conductor?**

- ○ a An oil service pipe
- ○ b A gas service pipe
- ○ c Flexible or pliable conduit
- ◉ d An extraneous-conductive-part

Answer d

See Part 5: Selection and erection of equipment, Regulations 543.2.1 and 543.2.2.

18 **Where a step-up transformer is used, all live conductors of the supply shall be disconnected from the supply by a**

- ○ a single-pole switch
- ○ b two-way switch
- ◉ c linked switch
- ○ d fused switch.

Answer c

See Part 5: Selection and erection of equipment, Regulation 555.1.3.

Notes

19 **Which of the following is the designated symbol for a high pressure sodium lamp with an internal starting device?**

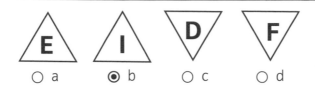

○ a   ⦿ b   ○ c   ○ d

Answer b

See Part 5: Selection and erection of equipment, Regulation 559.4 and Table 55.2.

Section 6

20 **Prospective fault current is recorded using the highest value determined from the prospective**

○ a   short circuits and earth leakage currents
⦿ b   short circuit and earth fault currents
○ c   earth leakage and breaking capacity currents
○ d   breaking capacity and earth fault currents.

Answer b

See Part 6: Inspection and testing, Regulation 612.11.

21 **For SELV and PELV circuits the separation of live parts from those of other circuits must be confirmed by**

○ a   inspection
⦿ b   measurement
○ c   calculation
○ d   enquiry.

Answer b

See Part 6: Inspection and testing, Regulations 612.4.1 and 612.4.2.

Section 7

22 Local supplementary bonding in a bathroom may be connected

- ○ a   near the origin of the installation at the consumer unit
- ◉ b   close to the entry of extraneous-conductive-parts to the room
- ○ c   to non-metallic water service pipes
- ○ d   to ceramic bath or shower basins.

Answer b

See Part 7: Special installations or locations, Regulation 701.415.2.

23 In zones 0 and 1 of a fountain, the protective measure of SELV may be used, provided that the source of SELV is

- ○ a   inside these zones
- ◉ b   outside these zones
- ○ c   via an autotransformer
- ○ d   centre tapped to earth.

Answer b

See Part 7: Special installations or locations, Regulation 702.410.3.4.2.

24 Which of the following supply systems is <u>not</u> to be used on construction sites, with the exception of a fixed building on the site?

- ◉ a   TN-C-S
- ○ b   TT
- ○ c   TN-S
- ○ d   Reduced low voltage

Answer a

See Part 7: Special installations or locations, Regulation 704.411.3.1.

25 **Enclosures used to protect electrical equipment vulnerable to external influences in a horticultural installation shall have a minimum degree of protection of**

- a   IP22
- b   IP33
- ⦿ c   IP44
- d   IP55.

Answer c

See Part 7: Special installations or locations, Regulation 705.512.2.

26 **The maximum length of a flexible cord or cable for a caravan connection is**

- a   1.8 m
- b   6 m
- c   10 m
- ⦿ d   25 m.

Answer d

See Part 7: Special installations or locations, Section 708, Fig 708, or Regulation 721.55.2.6.

27 **Any cable intended to supply temporary exhibition structures shall have, at its origin, an RCD having a maximum rated residual operating current of**

- a   30 mA
- b   100 mA
- ⦿ c   300 mA
- d   500 mA.

Answer c

See Part 7: Special installations or locations, Regulation 711.410.3.4.

28 Which of the following is regarded as a mobile or transportable unit?

- ○ a   Mobile generating sets
- ◉ b   Mobile medical services unit
- ○ c   Motor caravan
- ○ d   Mobile machinery

Answer b

See Part 7: Special installations or locations, Regulation 717.1.

Section 8

29 The installation reference method for single-core non-sheathed cables in trunking run horizontally on a wooden wall is

- ○ a   A
- ◉ b   B
- ○ c   C
- ○ d   E.

Answer b

See Appendices: Appendix 4, Table 4A2, Number 6.

30 The rating factor for a 70 °C thermoplastic cable installed in air in an ambient temperature of 40 °C is

- ○ a   1.0
- ○ b   0.91
- ◉ c   0.87
- ○ d   0.85.

Answer c

See Appendices: Appendix 4, Table 4B1.

Notes

Answer key

Sample test 1

Question	Answer
1	b
2	c
3	b
4	b
5	d
6	c
7	b
8	d
9	d
10	c
11	d
12	d
13	b
14	d
15	b
16	b
17	d
18	c
19	b
20	b
21	b
22	b
23	b
24	a
25	c
26	d
27	c
28	b
29	b
30	c

Exam practice 2

Sample test 2   46

Questions and answers   53

Answer key   62

Exam practice 2

Notes

Sample test 2

The sample test below has 30 questions, the same number as the online exam, and its structure follows that of the online exam. The test appears first without answers, so you can use it as a mock exam. It is then repeated with answers and references to the relevant section of the *IEE Wiring Regulations Seventeenth Edition*. Finally, there is an answer key for easy reference.

Answer the questions by filling in the circle next to your chosen option.

Section 1

1   BS 7671 does <u>not</u> apply to

○ a   equipment of aircraft
○ b   photovoltaic systems
○ c   marinas
○ d   fairgrounds.

2   Electrical installation design shall take into account

○ a   electromagnetic disturbances
○ b   direct lightning strikes
○ c   current world copper prices
○ d   local authority planning approval.

Section 2

3   Which of the following would be defined as a live part?

○ a   Earthing conductor
○ b   Line conductor
○ c   Exposed-conductive-part
○ d   PEN conductor

4   The symbol used to show that a BS 88 device has a motor circuit application is

○ a   gG
○ b   gM
○ c I_z
○ d I_2.

Section 3

5 The effectiveness of protective measures should be considered with regard to

- ○ a external influences
- ○ b safety services
- ○ c maintainability
- ○ d compatibility.

Section 4

6 The measure of automatic disconnection of supply is employed for a circuit supplying 13 A socket-outlets intended for general use by ordinary persons. Which of the following does <u>not</u> contribute to the provision of fault protection?

- ○ a Protective earthing
- ○ b Protective equipotential bonding
- ○ c Additional protection by RCD
- ○ d Reinforced insulation

7 Where U_o is 230 V and I_a is 100 A, the value of Z_s will be

- ○ a 0.43 Ω
- ○ b 2.3 Ω
- ○ c 23 Ω
- ○ d 0.023 MΩ.

8 The maximum value of earth fault loop impedance (Z_s) for a circuit protected by a 100 mA RCD forming part of a 230 V a.c. TT system is

- ○ a 500 Ω
- ○ b 460 Ω
- ○ c 167 Ω
- ○ d 92 Ω.

9 The maximum value of earth fault loop impedance (Z_s) for a 25 A BS EN 60898 type D circuit-breaker protecting a 110 V single-phase reduced voltage circuit is

- ○ a 0.44 Ω
- ○ b 0.26 Ω
- ○ c 0.22 Ω
- ○ d 0.11 Ω.

Notes

10 Where additional protection is provided to basic protection, a 30 mA RCD shall disconnect within 40 ms at a residual current of

○ a   15 mA
○ b   30 mA
○ c   100 mA
○ d   150 mA.

11 Basic protection may be provided by

○ a   barriers and enclosures to IPXXB or IP2X
○ b   fuses and circuit-breakers
○ c   supplementary equipotential bonding
○ d   backup protection.

12 Which of the following need <u>not</u> be tested under fire conditions to ensure compliance with non-flame propagating requirements?

○ a   Cables
○ b   Protective devices
○ c   Conduit systems
○ d   Trunking systems

13 In the event of an earth fault on the HV side of a substation the LV installation may be affected by

○ a U_o
○ b U_f
○ c I_d
○ d I^2t.

Section 5

14 A cable concealed in a wall outside the prescribed zones at a depth of less than 50 mm must

○ a   not be installed
○ b   be enclosed in unearthed conduit
○ c   be enclosed in earthed metallic conduit
○ d   be protected by a 500 mA RCD.

15 Circuits supplied by a generator set which is not permanently fixed shall have additional protection by a

- ○ a   BS 88 device only
- ○ b   30 mA RCD
- ○ c   100 mA RCD
- ○ d   300 mA RCD.

16 The equipment, design, installation and testing of an electric surface heating system shall be in accordance with

- ○ a   BS EN 60898
- ○ b   BS 6217
- ○ c   BS 6351
- ○ d   BS EN 60417.

17 Lighting circuits incorporating E40 lampholders shall be protected by an overcurrent device with a maximum rating of

- ○ a   6 A
- ○ b   10 A
- ○ c   16 A
- ○ d   20 A.

18 Which of the following is the symbol for a "class P" thermally protected independent lamp ballast permitted for mounting on a flammable surface?

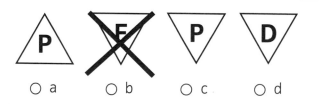

    ○ a   ○ b   ○ c   ○ d

19 A wiring system is to be installed between a safety source and a main distribution board. The risks required to be reduced to a minimum do not include

- ○ a   short-circuit
- ○ b   earth fault
- ○ c   ageing
- ○ d   fire.

Section 6

20 A simple method to allow for measured values of loop impedance to be effectively compared with tabulated maximum values is to correct these maximum values by multiplying them by

- ○ a 0.75
- ○ b 0.8
- ○ c 1.2
- ○ d 1.8.

21 The test voltage and minimum insulation resistance value for a PELV circuit is

- ○ a 250 V, 1 Ω
- ○ b 500 V, 0.5 MΩ
- ○ c 230 V, 1 MΩ
- ○ d 250 V, 0.5 MΩ.

Section 7

22 13 A socket-outlets are to be installed in a location containing a bathtub. The minimum distance they may be located from zone 1 is

- ○ a 1 m
- ○ b 2 m
- ○ c 3 m
- ○ d 4 m.

23 Additional protection for all lighting circuits within a sauna shall be provided by

- ○ a an insulation monitoring device
- ○ b an RCD with a rated residual operating current not exceeding 30 mA
- ○ c barriers and enclosures having a degree of protection of at least IP54
- ○ d ensuring the luminaires have a degree of protection of at least IPX3.

24 For an agricultural location, in order to provide automatic disconnection of supply in circuits other than socket-outlet circuits, a disconnection device shall be installed having a maximum rated residual operating current of

- ○ a   30 mA
- ○ b   100 mA
- ○ c   300 mA
- ○ d   500 mA.

25 Within a conducting location with restricted movement, supplies to 110 V mobile equipment must provide protection against electric shock by the use of

- ○ a   electrical separation
- ○ b   Class II protection
- ○ c   obstacles
- ○ d   PELV.

26 Equipment installed on a pontoon located in a marina, which is likely to be subjected to splashes, shall have a degree of protection at least

- ○ a   IPX3
- ○ b   IPX4
- ○ c   IPX5
- ○ d   IPX6.

27 When the a.c. side of photovoltaic equipment is disconnected, the d.c. side is considered to be

- ○ a   de-energised
- ○ b   energised
- ○ c   à non-conducting location
- ○ d   an equipotential zone.

28 Lighting chains used in a circus may be any length provided that

- ○ a   undervoltage protection is present
- ○ b   overcurrent devices are properly rated
- ○ c   supplementary bonding is present
- ○ d   an external influence of AD4 is not exceeded.

Notes

Section 8

29 A flat twin and earth cable clipped direct to a ceiling joist where the thermal insulation does not exceed 100 mm thickness is installation method

○ a   100
○ b   101
○ c   102
○ d   103.

30 The maximum value of voltage drop for lighting in a low voltage installation supplied from a public distribution system is

○ a   3%
○ b   4%
○ c   5%
○ d   6%.

Questions and answers

The questions and answers in sample test 2 are repeated below with answers and references to the relevant section of the *IEE Wiring Regulations Seventeenth Edition*.

Section 1

1   BS 7671 does <u>not</u> apply to

- ◉ a   equipment of aircraft
- ○ b   photovoltaic systems
- ○ c   marinas
- ○ d   fairgrounds.

Answer a
See Part 1: Scope, Regulation 110.2.

2   Electrical installation design shall take into account

- ◉ a   electromagnetic disturbances
- ○ b   direct lightning strikes
- ○ c   current world copper prices
- ○ d   local authority planning approval.

Answer a
See Part 1: Scope, Regulation 131.6.4.

Section 2

3   Which of the following would be defined as a live part?

- ○ a   Earthing conductor
- ◉ b   Line conductor
- ○ c   Exposed-conductive-part
- ○ d   PEN conductor

Answer b
See Part 2: Definitions.

Notes

4   **The symbol used to show that a BS 88 device has a motor circuit application is**

○ a   gG
◉ b   gM
○ c I_Z
○ d I_2.

Answer b
See Part 2: Definitions, Symbols.

Section 3

5   **The effectiveness of protective measures should be considered with regard to**

○ a   external influences
○ b   safety services
◉ c   maintainability
○ d   compatibility.

Answer c
See Part 3: Assessment of general characteristics, Regulation 341.1.

Section 4

6   **The measure of automatic disconnection of supply is employed for a circuit supplying 13 A socket-outlets intended for general use by ordinary persons. Which of the following does <u>not</u> contribute to the provision of fault protection?**

○ a   Protective earthing
○ b   Protective equipotential bonding
○ c   Additional protection by RCD
◉ d   Reinforced insulation

Answer d
See Part 4: Protection for safety, Regulation 411.1 (ii).

7 Where U_o is 230 V and I_a is 100 A, the value of Z_s will be

○ a   0.43 Ω
◉ b   2.3 Ω
○ c   23 Ω
○ d   0.023 MΩ.

Answer b
See Part 4: Protection for safety, Regulation 411.4.5.

8 The maximum value of earth fault loop impedance (Z_s) for a circuit protected by a 100 mA RCD forming part of a 230 V a.c. TT system is

◉ a   500 Ω
○ b   460 Ω
○ c   167 Ω
○ d   92 Ω.

Answer a
See Part 4: Protection for safety, Regulation 411.5.3, Table 41.5.

9 The maximum value of earth fault loop impedance (Z_s) for a 25 A BS EN 60898 type D circuit-breaker protecting a 110 V single-phase reduced voltage circuit is

○ a   0.44 Ω
○ b   0.26 Ω
○ c   0.22 Ω
◉ d   0.11 Ω.

Answer d
See Part 4: Protection for safety, Regulation 411.8.3, Table 41.6.

10 Where additional protection is provided to basic protection, a 30 mA RCD shall disconnect within 40 ms at a residual current of

○ a   15 mA
○ b   30 mA
○ c   100 mA
◉ d   150 mA.

Answer d
See Part 4: Protection for safety, Regulation 415.1.1.

11 Basic protection may be provided by

- ⊙ a barriers and enclosures to IPXXB or IP2X
- ○ b fuses and circuit-breakers
- ○ c supplementary equipotential bonding
- ○ d backup protection.

Answer a

See Part 4: Protection for safety, Regulation 416.2.1.

12 Which of the following need <u>not</u> be tested under fire conditions to ensure compliance with non-flame propagating requirements?

- ○ a Cables
- ⊙ b Protective devices
- ○ c Conduit systems
- ○ d Trunking systems

Answer b

See Part 4: Protection for safety, Regulation 422.2.1.

13 In the event of an earth fault on the HV side of a substation the LV installation may be affected by

- ○ a U_o
- ⊙ b U_f
- ○ c I_d
- ○ d I^2t.

Answer b

See Part 4: Protection for safety, Regulation 442.2.

Section 5

14 A cable concealed in a wall outside the prescribed zones at a depth of less than 50 mm must

- ○ a not be installed
- ○ b be enclosed in unearthed conduit
- ⊙ c be enclosed in earthed metallic conduit
- ○ d be protected by a 500 mA RCD.

Answer c

See Part 5: Selection and erection of equipment, Regulation 522.6.6.

Notes

15 Circuits supplied by a generator set which is not permanently fixed shall have additional protection by a

- ○ a   BS 88 device only
- ◉ b   30 mA RCD
- ○ c   100 mA RCD
- ○ d   300 mA RCD.

Answer b
See Part 5: Selection and erection of equipment, Regulation 551.4.4.2.

16 The equipment, design, installation and testing of an electric surface heating system shall be in accordance with

- ○ a   BS EN 60898
- ○ b   BS 6217
- ◉ c   BS 6351
- ○ d   BS EN 60417.

Answer c
See Part 5: Selection and erection of equipment, Regulation 554.5.1.

17 Lighting circuits incorporating E40 lampholders shall be protected by an overcurrent device with a maximum rating of

- ○ a   6 A
- ○ b   10 A
- ◉ c   16 A
- ○ d   20 A.

Answer c
See Part 5: Selection and erection of equipment, Regulation 559.6.1.6.

Notes

18 Which of the following is the symbol for a "class P" thermally protected independent lamp ballast permitted for mounting on a flammable surface?

   ○ a   ○ b   ◉ c   ○ d

Answer c
See Part 5: Selection and erection of equipment, Regulation 559.7 and Table 55.2.

19 A wiring system is to be installed between a safety source and a main distribution board. The risks required to be reduced to a minimum do **not** include

○ a   short-circuit
○ b   earth fault
◉ c   ageing
○ d   fire.

Answer c
See Part 5: Selection and erection of equipment, Regulation 560.8.3.

Section 6

20 A simple method to allow for measured values of loop impedance to be effectively compared with tabulated maximum values is to correct these maximum values by multiplying them by

○ a   0.75
◉ b   0.8
○ c   1.2
○ d   1.8.

Answer b
See Part 6: Inspection and testing, Regulation 612.9, and Appendix 14.

21 **The test voltage and minimum insulation resistance value for a PELV circuit is**

○ a   250 V, 1 Ω
○ b   500 V, 0.5 MΩ
○ c· 230 V, 1 MΩ
◉ d   250 V, 0.5 MΩ.

Answer d
See Part 6: Inspection and testing, Regulation 612.3.2, Table 61.

Section 7

22 13 **A socket-outlets are to be installed in a location containing a bathtub. The minimum distance they may be located from zone 1 is**

○ a   1 m
○ b   2 m
◉ c   3 m
○ d   4 m.

Answer c
See Part 7: Special installations or locations, Regulation 701.512.3.

23 **Additional protection for all lighting circuits within a sauna shall be provided by**

○ a   an insulation monitoring device
◉ b   an RCD with a rated residual operating current not exceeding 30 mA
○ c   barriers and enclosures having a degree of protection of at least IP54
○ d   ensuring the luminaires have a degree of protection of at least IPX3.

Answer b
See Part 7: Special installations or locations, Regulation 703.411.3.3.

Notes

24 **For an agricultural location, in order to provide automatic disconnection of supply in circuits other than socket-outlet circuits, a disconnection device shall be installed having a maximum rated residual operating current of**

- ○ a   30 mA
- ○ b   100 mA
- ◉ c   300 mA
- ○ d   500 mA.

Answer c

See Part 7: Special installations or locations, Regulation 705.411.1.

25 **Within a conducting location with restricted movement, supplies to 110 V mobile equipment must provide protection against electric shock by the use of**

- ◉ a   electrical separation
- ○ b   Class II protection
- ○ c   obstacles
- ○ d   PELV.

Answer a

See Part 7: Special installations or locations, Regulation 706.410.3.10.

26 **Equipment installed on a pontoon located in a marina, which is likely to be subjected to splashes, shall have a degree of protection at least**

- ○ a   IPX3
- ◉ b   IPX4
- ○ c   IPX5
- ○ d   IPX6.

Answer b

See Part 7: Special installations or locations, Regulation 709.512.2.1.1.

27 When the a.c. side of photovoltaic equipment is disconnected, the d.c. side is considered to be

- ○ a   de-energised
- ◉ b   energised
- ○ c   a non-conducting location
- ○ d   an equipotential zone.

Answer b
See Part 7: Special installations or locations, Regulation 712.410.3.

28 Lighting chains used in a circus may be any length provided that

- ○ a   undervoltage protection is present
- ◉ b   overcurrent devices are properly rated
- ○ c   supplementary bonding is present
- ○ d   an external influence of AD4 is not exceeded.

Answer b
See Part 7: Special installations or locations, Regulation 740.55.1.1, note.

Section 8

29 A flat twin and earth cable clipped direct to a ceiling joist where the thermal insulation does not exceed 100 mm thickness is installation method

- ◉ a   100
- ○ b   101
- ○ c   102
- ○ d   103.

Answer a
See Appendices: Appendix 4, Table 4A2.

30 The maximum value of voltage drop for lighting in a low voltage installation supplied from a public distribution system is

- ◉ a   3%
- ○ b   4%
- ○ c   5%
- ○ d   6%.

Answer a
See Appendices: Appendix 12, Table 12A.

Notes

Answer key

Sample test 2

Question	Answer
1	a
2	a
3	b
4	b
5	c
6	d
7	b
8	a
9	d
10	d
11	a
12	b
13	b
14	c
15	b
16	c
17	c
18	c
19	c
20	b
21	d
22	c
23	b
24	c
25	a
26	b
27	b
28	b
29	a
30	a

More information

Further reading   64

Online resources   65

Further courses   66

More information

Notes

Further reading

Required reading
BS 7671: 2008 Requirements for Electrical Installations, IEE Wiring Regulations Seventeenth Edition, published by the IEE, London, 2008

Additional reading
On-Site Guide: BS 7671:2008, published by the IEE, London, 2008

The Electrician's Guide to Good Electrical Practice, published by Amicus, 2005

Electrician's Guide to the Building Regulations, published by the IEE, London, 2005

IEE Guidance Notes, a series of guidance notes, each of which enlarges upon and amplifies the particular requirements of a part of the Wiring Regulations, 17th Edition, published by the IEE, London:
– Guidance Note 1: *Selection and Erection of Equipment*, 5th Edition 2008
– Guidance Note 2: *Isolation and Switching*, 5th Edition 2008
– Guidance Note 3: *Inspection and Testing*, 5th Edition 2008
– Guidance Note 4: *Protection Against Fire*, 5th Edition 2008
– Guidance Note 5: *Protection Against Electric Shock*, 5th Edition 2008
– Guidance Note 6: *Protection Against Overcurrent*, 5th Edition 2008
– Guidance Note 7: *Special Locations*, 3rd Edition 2008

Brian Scaddan, *Electrical Installation Work*, published by Newnes (an imprint of Butterworth-Heinemann), 2002

John Whitfield, *Electrical Craft Principles*, published by the IEE, London, 1995

Online resources

Notes

City & Guilds www.cityandguilds.com
The City & Guilds website can give you more information about studying for further professional and vocational qualifications to advance your personal or career development, as well as locations of centres that provide the courses.

Institution of Engineering and Technology (IET) www.theiet.org
The Institution of Engineering and Technology was formed by the amalgamation of the Institution of Electrical Engineers (IEE) and the Institution of Incorporated Engineers (IIE). It is the largest professional engineering society in Europe and the second largest of its type in the world. The Institution produces the *IEE Wiring Regulations* and a range of supporting material and courses.

SmartScreen www.smartscreen.co.uk
City & Guilds' dedicated online support portal SmartScreen provides learner and tutor support for over 100 City & Guilds qualifications. It helps engage learners in the excitement of learning and enables tutors to free up more time to do what they love the most – teach!

BRE Certification Ltd www.partp.co.uk

British Standards Institution www.bsi-global.com

CORGI Services Ltd www.corgi-gas-safety.com

ELECSA Ltd www.elecsa.org.uk

Electrical Contractors' Association (ECA) www.eca.co.uk

Joint Industry Board for the Electrical Contracting Industry (JIB)
www.jib.org.uk

NAPIT Certification Services Ltd www.napit.org.uk

National Inspection Council for Electrical Installation Contracting (NICEIC) www.niceic.org.uk

Oil Firing Technical Association for the Petroleum Industry (OFTEC)
www.oftec.co.uk

Notes

Further courses

City & Guilds Level 3 Certificate in Inspection, Testing and Certification of Electrical Installations (2391-10)

This course is aimed at those with practical experience of inspection and testing of LV electrical installations, who require to become certificated possibly for NICEIC purposes. It is not suitable for beginners. In addition to relevant practical experience, candidates must possess a good working knowledge of the requirements of BS 7671 to City & Guilds Level 3 certificate standard or equivalent.

City & Guilds Level 3 Certificate in the Code of Practice for In-Service Inspection and Testing of Electrical Equipment (2377)

This course, commonly known as PAT/Portable Appliance Testing, is for staff undertaking and recording inspection and testing of electrical equipment. The course includes a practical exercise. Topics covered: equipment construction, inspection and recording, combined inspection and testing, and equipment.

City & Guilds Level 2 Certificate in Fundamental Inspection, Testing and Initial Verification (2392-10)

This qualification was developed to meet industry needs and to provide candidates with an introduction to the initial verification of electrical installations. It is aimed at practising electricians who have not carried out inspection and testing since qualifying, those who require update training and those with limited experience of inspection and testing. Together with suitable on-site experience, it would also prepare candidates to go on to the Level 3 Certificate in Inspection, Testing and Certification of Electrical Installations (2391-10).

City & Guilds Building Regulations for Electrical Safety

This new suite of qualifications is for Competent Persons in Domestic Electrical Installations (Part P of the Building Regulations). The qualifications consist of components for specialised domestic building regulations and domestic wiring regulations routes as well as a component for Qualified Supervisors.

JIB Electrotechnical Certification Scheme (ECS) Health and Safety Assessment

This Health and Safety Assessment is a requirement for electricians wishing to work on larger construction projects and sites in the UK and the exam is an online type very similar in format to the GOLA tests. It is now a mandatory requirement for holding an ECS card, and is a requirement for all members of the ECS. Please refer to www.jib.org.uk/ecs2.htm for details.

Notes

Notes